EXCEL

FORMULAS AND FUNCTIONS

For Complete Beginners, Step-By-Step
Illustrated Guide to Mastering
Formulas and Functions

Table of Contents

Introduction

Hello, there future Excel Programmers! Thanks for viewing this book. This book has been designed to be your go-to book for excel programming. Spreadsheets have been with us for a long time. The best-known and most widely used spreadsheet is Microsoft Excel. Excel is easy to use for most daily number crunching tasks and it comes with formulas and functions that perform a host of tasks. Tasks such as summing, manipulating text, averaging, comparing, answering what if questions...

Assuming you know the meaning of such basic spreadsheet terms, as row, column, and cell, and you know how to do such basic operations as copy and paste, this little book will introduce you to a number of the formulas and

functions of Excel. It will give you the
ability and confidence to use them
successfully.

Chapter 1: What are formulas and functions?

A **formula** is an expression used to calculate the value in a cell.

For example, here is a very simple problem where A2 contains 5.7 and B2 contains 6.4. The task is to put the products of these two numbers in into cell C2. See diagram below.

A	B	C
5.7	6.4	

We can do this by selecting c2 and writing the very simple formula = **A2*B2** then pressing enter [it does not matter if you write **a2*B2**, **A2*b2** or **a2*b2**, as uppercase and lower case letters are treated the same when referring to columns].

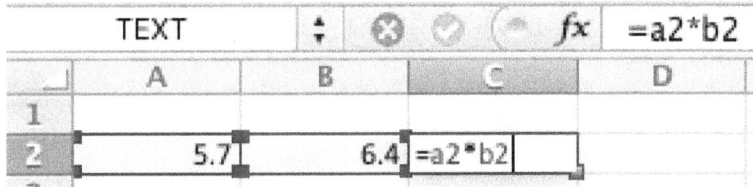

	A	B	C	D
1				
2	5.7	6.4	=a2*b2	

When this is done it leads to the value of the product appearing in the cell C2. [The word 'TEXT' and the expression fx = A2*B2, which are shown in the diagram appear automatically. You can ignore them.]

2	5.7	6.4	36.48

Another simple example is adding the contents of the cells A2, A3, A4, and putting the results in A5.

	A
1	
2	5.7
3	7.8
4	10.3
5	

Again, we select A5 and write =
A2+A3+A4.

	A
1	
2	5.7
3	7.8
4	10.3
5	=a2+a3+a4

After we press enter the result is
obtained and ends up in A5.

	A
1	
2	5.7
3	7.8
4	10.3
5	23.8

So far so good, but what if you needed to
add the contents of 100 cells. It would
take a long time to write =
A2+A3+A4+ A5...... +A100 + A101
and you would probably make an error.

97	51
98	28
99	94
100	52
101	34

Fortunately, the creators of Excel anticipated many needs of users decades ago and created functions. For this particular problem, we use the sum function.

Select cell c102 and type, '= **SUM(a2:a101)** '[this is called the function declaration].

97	51
98	28
99	94
100	52
101	34
02	=SUM(a2:a101)

then press enter and cell a102 has the sum, which in this case was 6086.

97	51
98	28
99	94
100	52
101	34
102	6086

A **function** is a predefined formula, which performs calculations with the contents of cells.

Parameters

You may notice in the expression, '=SUM(a2:a101**)'** the interesting phrase a2:a101. This is the *range* of the function. The range of a function is the set of cells on which it acts. The range of the SUM function is called its *parameter*, as it can vary. We will have different ranges to which the SUM function is applied.

There are a vast number of functions in Excel, which you will become familiar with as you master the use of this amazing tool. This little book will look at some of the more important functions.

Chapter 2: Text Formulas

The Excel **TEXT** formula or function is a most interesting one.

Date format

The Excel TEXT formula or function can be used to bring about changes of date and number format. Here is an example, suppose we wanted to change a column of dates in a certain format

2		19-Mar-01
3		20-Mar-01
4		21-Mar-01
5		22-Mar-01
6		23-Mar-01
7		24-Mar-01
8		25-Mar-01

and wish to change them into the form mm/dd/yyyy. In order to do this, we use the Text function.

We can pick any column of 7 cells to put them in, so let's put them in the column starting at a2. Select a2 and type the function declaration,'

=TEXT(b2,"mm/dd/yyyy") '

TEXT		fx	=TEXT(b2,"mm/dd/yyyy")		
	A	B	C	D	E	F
1						
2	=TEXT(b2,"mm/dd/yyyy")					
3	TEXT(value, format_text)					
4		21-Mar-01				
5		22-Mar-01				
6		23-Mar-01				
7		24-Mar-01				
8		25-Mar-01				

If you do this, Excel automatically offers you possible functions to make your task easier. You are sensible to use them. Once you filled the function in press enter and you will find the correct expression in a2.

You can quickly fill in the rest of the column by using the 'handle' at the side of a2.

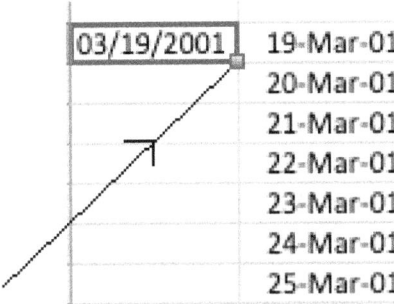

03/19/2001	19-Mar-01
	20-Mar-01
	21-Mar-01
	22-Mar-01
	23-Mar-01
	24-Mar-01
	25-Mar-01

Then drag down.

2	03/19/2001	19-Mar-01
3	03/20/2001	20-Mar-01
4	03/21/2001	21-Mar-01
5	03/22/2001	22-Mar-01
6	03/23/2001	23-Mar-01
7	03/24/2001	24-Mar-01
8	03/25/2001	25-Mar-01

Currency

Another possible use of the **TEXT** function is to convert a column of numbers

E
5
6
78.9
23.97

into currency format.

Note the first number is contained in e2.
Suppose the new column will start at
f10. Select f10 and type the function
declaration, ' = **TEXT(e2,"$##.##")** '

```
=TEXT(e2,"$##.##")
    TEXT(value, format_text)
```

Press enter.

Now use the handle and move down to
get the remaining conversions.

$5.
$6.
$78.9
$23.97

TEXT can be used in many ways like the
two examples shown.

If we look at the structure or syntax of the TEXT function through a declaration, '=**TEXT(e2,"\$##.##")**', we see the parameters of the TEXT function are TEXT(value, format_text). Value, as the contents of cells, is values; format_text, as the function acts to change the format of the value.

Chapter 3: Comparison Formulas

Sometimes you wish to compare two columns.

87	64
57	28
92	23
81	33
64	14
56	91

You test whether two specific cells match with the IF function [We will say more about this later]. Begin a formula with =**IF(** and enter the two cell locations, with = between them. Put in a comma then enter relevant text, in quotes, to show if the cells match. Put in another comma and then the relevant text, in quotes, if there is a non-match.

87	64	=IF(I11=J11,"Match","No Match")
57	28	IF(logical_test, [value_if_true], [value_if_false])
92	23	
81	33	
64	14	
56	91	

Now press **Enter** to complete the

formula.

87	64	**No Match**
57	28	
92	23	
81	33	
64	14	
56	91	

Use the handle to complete the

comparison.

87	64	**No Match**
57	28	**No Match**
92	23	**No Match**
81	33	**No Match**
64	14	**No Match**
56	91	**No Match**

Chapter 4: Operators

Excel has operators in four categories. These are arithmetic operators, comparison operators, text concatenation operators, and reference operators.

The arithmetic operators are the usual +, -, × and ÷ operators except that in Excel × is written as * and ÷ as/. There are two other important arithmetic operators. These are % [percentage, which calculates a percentage] and ^ [caret, this raises to powers. See below].

3	2	=b2^c2

Press enter

3	2	9

The comparison operators are as they are in arithmetic and they are: >, < , ≤ , ≥ , =, ≠, however Excel uses <= for ≤ , >= for ≥ and <> for ≠.

Text concatenation may be a new idea for some of you. There is only one text concatenation operator and that is & [ampersand]. Here is how it works, suppose you had the string [word], "nice" in a2 and "cat" in b2

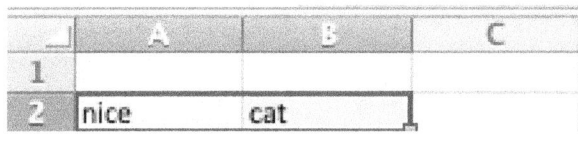

then = **a2 & b2** gives "nicecat".

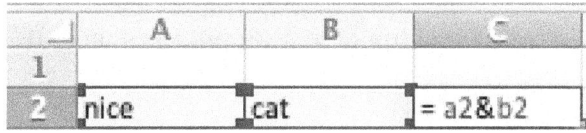

	A	B	C
1			
2	nice	cat	nicecat

All this operator does is join words or strings into one word or string.

Another type of operator, which may be new to you is the type referred to as a reference operator. Here are the reference operators.

: is the colon operator and refers to all references in a range from one cell to another including the endpoints.

Example. a2:a4 refers to a2, a3, and a4.

, is the comma operator and joins two ranges together. Here is an example a2: a4, b2: b4 refers to a2, a3, a4, b2, b3, b4. It is the equivalent in mathematics of the union operator [∪].

Finally, we have the single space operator. This is just an empty space. It produces references to common cells and ranges.

Example. a2:a4 a3:a5 gives a3, a4. Obviously, a3 and a4 are the only common cells of the ranges a2, a3, a4 and a3, a4, a5. It is the equivalent in mathematics of the intersection operator [∩].

A very important consideration in constructing formulas is the precedence or order of operators, in order to get the

correct result. All formulas begin with =
and if you are dealing with numbers the
order is similar to the order of
arithmetical operators as described by
BEDMAS [Brackets→Exponents→
Division and Multiplication→Addition
and Subtraction] or PEMDAS
[Brackets→Exponents→ Multiplication
and Division →Addition and
Subtraction] taught in high school.

Here is how the precedence of Excel
operations works.

: [colon], single-space, , [comma],
negation [as in -7], % [percentage], ^
[caret], * and/[neither is above the
other], + and - /[neither is above the
other], & [ampersand], =, <,>,<=,
>=,<> [comparison].

Chapter 5: Absolute vs Relative Cell References

When you working with Excel, you must know about what is called relative vs. absolute cell reference.

The problem is this: if you COPY A FORMULA containing cell references, generally the CELL REFERENCES CHANGE!

The diagrams below illustrate this.

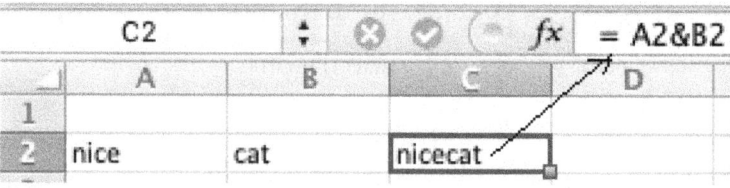

Now copy these three cells to a1,b1,c1
and look what happens.

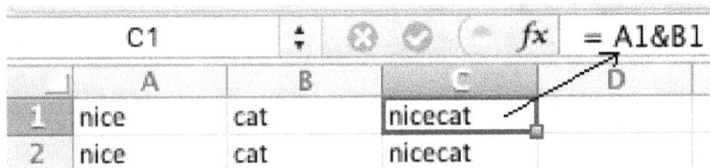

These cell references are named
"relative" cell references, as they change
relative to where the formula is copied.

Sometimes, you don't want cell
references altering when a formula is
copied. If that is the case then you use
what is called absolute cell references. In
order to use absolute cell references, you
put a "$" before the column letter if you
want that to always remain the same.
Similarly, you put a "$" in front of the
row number if you want that to remain
the same.

The following diagrams should help you grasp this very important idea.

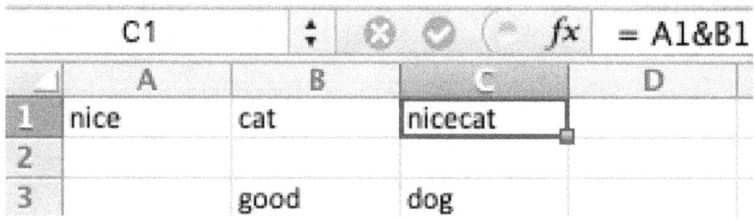

In this diagram, we are using relative references. Now copy the cell C2 to D3.

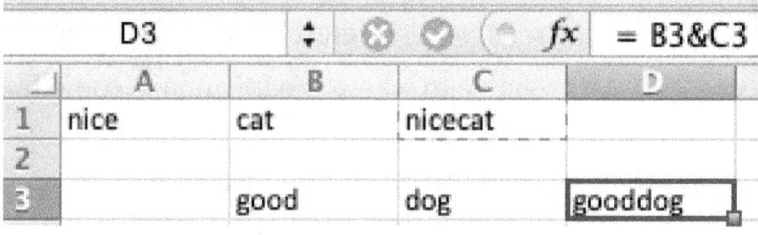

Note how the cells have changed from A1→B3 and B1→C3.

Now let's do this using Absolute References.

C1		⊗ ⊘	fx	= $A$1&$B$1	
	A	B	C	D	E
1	nice	cat	nicecat		
2					
3		good	dog		

The first diagram is as we got last time because "nice" is actually in the cell A1 and "cat" is actually in B1. Now let's copy the cell C2 to D3, as we did before

D3		⊗ ⊘	fx	= $A$1&$B$1	
	A	B	C	D	E
1	nice	cat	nicecat		
2					
3		good	dog	nicecat	

This time we got 'nicecat' instead of 'gooddog' because the reference was to A1 and B1 exactly and not just to the two cells immediately to the left of the cell we were copying.

When you create spreadsheets decide before copying a formula the cell references that are relative and those that are absolute.

You will realize the absolute importance of these ideas when we look at the VLOOKUP and HLOOKUP functions later.

Chapter 6: SUM

We had a look at the SUM function in chapter 1. If you have forgotten this have another read. However, here is another look at this extraordinary function.

This time, we're going to sum 3 columns of numbers.

A	B	C	D
	66	57	42
	64	77	49
	28	71	85
	23	59	61
	33	47	28
	14	50	88
	91	57	41
	87	53	88
	57	78	71
	92	81	16

In order to do this, we shall use the : and , operators discussed in the last chapter. In D12, write,' **=SUM(b2:b11,c2:c11, d2:d11)** ' then press enter. The sum of these numbers is 1754.

	D12		⊗ ⊘	fx	=SUM(B2:B11,C2:C11,D2:D11)
	A	B	C	D	E	F
1						
2		66	57	42		
3		64	77	49		
4		28	71	85		
5		23	59	61		
6		33	47	28		
7		14	50	88		
8		91	57	41		
9		87	53	88		
10		57	78	71		
11		92	81	16		
12				1754		

Chapter 7: IF

In many ways, the IF function is the most important function Excel has. Earlier, we showed how it could be used for comparisons. Here is another example of this. We are going to have two columns of numbers. Using IF, a column to the right will have the word "CAT" if the number in the left column is no less than the number in the right. If this condition is not met, in other words, if the number in the left column is less than the number in the right then we will get "DOG". Here are the columns.

	A	B	C
1			
2		88	57
3		42	51
4		57	58
5		77	42
6		71	49
7		59	85
8		47	61
9		50	28
10		57	88
11		53	41
12		78	88
13		81	71
14		81	16
15		61	56
16		53	59

Into cell a2 type the function declaration,' = **IF(b2>=c2, "CAT","DOG")** ' then press enter.

D2				fx	=IF(B2>=C2,"CAT","DOG")		
	A	B	C	D	E	F	
1							
2		88	57	CAT			
3		42	51				

Now, you can use the 'handle' to complete the use of the function.

	A	B	C	D	E	F
D4				fx	=IF(B4>=C4,"CAT","DOG")	
1						
2		88	57	CAT		
3		42	51	DOG		
4		57	58	DOG		
5		77	42	CAT		

Now before leaving the IF function, we will consider another extremely useful property of this extraordinary function:- NESTING. In this context, nesting has nothing to do with birds. It is the ability of the IF function to make itself one of the choices. Here is an example to show this.

Example. We are given the following column of numbers.

	A	B	C
1			
2		188	
3		1142	
4		57	
5		77	
6		1	
7		459	
8		47	
9		504	
0		257	
1		6753	
2		798	
3		81	
4		281	
5		61	
6		5	

Into C column beside the numbers in B column, there will be sentences saying," Bi has a number with x digits." Thus by 188, there will be, "188 has 3 digits. The next diagram will show how this is done using the IF function.

C2			fx =IF(B2>999,B2 & " has 4 digits",IF(B2>99,B2 & " has 3 digits",IF(B2>9,B2 & "has 2 digits",B2 & " has1 digi
	A	B	D E F G H I J K L
1			
2		188	188 has 3 digits
3		1142	
4		57	
5		77	
6		1	
7		459	
8		47	
9		504	
10		257	
11		6753	
12		798	
13		81	
14		281	
15		61	
16		5	

Using the 'handles' solves the final part of the problem.

188	188 has 3 digits
1142	1142 has 4 digits
57	57has 2 digits
77	77has 2 digits
1	1 has1 digit
459	459 has 3 digits
47	47has 2 digits
504	504 has 3 digits
257	257 has 3 digits
6753	6753 has 4 digits
798	798 has 3 digits
81	81has 2 digits
281	281 has 3 digits
61	61has 2 digits
5	5 has1 digit

Structure or Syntax of IF

If we look at the structure or syntax of the IF function through a declaration, ' **= IF(b2>=c2, "CAT","DOG")** ', we see the parameters of the IF function are IF(logical test, result if TRUE, result if FALSE). Logical test, which is either TRUE or FALSE, as the contents of cells are values; Result if TRUE; Result if FALSE.

Chapter 8: AND

Another widely used Excel function is the AND function. It is either TRUE or FALSE depending on whether two conditions are both true or not.

Here is an example, which illustrates this. We need to compare the sizes of the numbers contained in the rows of this table using the AND function.

	A	B
1		
2	23	90
3	56	78
4	34	48
5	55	12

C10

Into cell C2 type the function declaration,' **=AND(A2<50,B2<50)**'

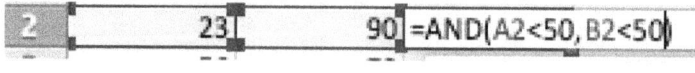

then press enter.

	C2			⊗ ✓	fx	=AND(A2<50,B2<50)

	A	B	C	D	E
1					
2	23	90	FALSE		
3	56	78			
4	34	48			
5	55	12			

Now use the handle to complete the task

2	23	90	FALSE
3	56	78	FALSE
4	34	48	TRUE
5	55	12	FALSE

The only row, which had A and B cells both less than 50 was row 4.

Here is another example of the same type. Find the rows in which the C and D cells are both less than 50 in the table below.

C	D
14	57
12	22
87	71
57	9
92	47
16	50
64	57
56	53
17	22
46	81
88	23

Into cell E2 type the function declaration, ' =AND(C2<50,D2<50) ' then press enter.

fx	=AND(C2<50,D2<50)	

C	D	E
14	57	FALSE
12	22	
87	71	
57	9	
92	47	
16	50	
64	57	
56	53	
17	22	
46	81	
88	23	

Now use the handle to complete the task.

14	57	FALSE
12	22	TRUE
87	71	FALSE
57	9	FALSE
92	47	FALSE
16	50	FALSE
64	57	FALSE
56	53	FALSE
17	22	TRUE
46	81	FALSE
88	23	FALSE

In this case, only rows 3 and 10 had C and D cells where the number was less than 50.

Structure or Syntax of AND

If we look at the structure or syntax of the AND function through a declaration, **' =AND(C2<50, D2<50)'**, we see the parameters of the AND function are AND(logical test, logical test). The parameters are two logical tests, the combination of whose results give different results.

Chapter 9: LEN

The LEN function is one of a large number of Excel functions, which handle strings [words]. The LEN function has a very simple format = **LEN (**string**)**. The function returns the number of letters in a string, which is often in a cell.

Here is an example. We have a column of names and need to write beside it the number of letters in each of the names.

	A
1	
2	Jimmy
3	Jamie
4	Marsden
5	Oliver
6	Sean
7	Jaxon
8	Bart
9	Daniel
10	Dominic
11	Alex
12	Leo
13	Vinu
14	Harry
15	Vincent
16	Oliver
17	Nathan
18	Jean-Michael
19	Scott
20	Matthew
21	Siddhartha

The input in cell B2 is = LEN [A2].

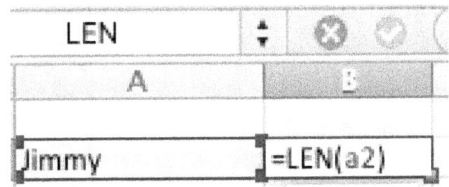

A	B
Jimmy	=LEN(a2)

Now press enter. This action produces

B2				fx	=LEN(A2)
	A	B	C		D
1					
2	Jimmy	5			

Then we use the handle to produce

D21	
A	B
Jimmy	5
Jamie	5
Marsden	7
Oliver	6
Sean	4
Jaxon	5
Bart	4
Daniel	6
Dominic	7
Alex	4
Leo	3
Vinu	4
Harry	5
Vincent	7
Oliver	6
Nathan	6
Jean-Michael	12
Scott	5
Matthew	7
Siddhartha	10

There are many string functions in Excel. With Excel string functions, you can do all sorts of things, such as joining text from different cells to make a new one string, removing parts of a string depending on their position or surroundings, substitution into parts of a string, etc. Sometimes, these functions are referred to as text functions. This sometimes leads to confusion with the TEXT function, which we looked at in Chapter 2 .

The LEN function is a string function, some other very important string functions are the LEFT and Right Functions. The LEFT function gets a substring that contains a given *number of left characters* from a string. Similarly, the RIGHT function gets a substring that contains a given *number of right characters* from a string. The following example shows how the LEFT function works.

Example. There is a column of names from which we want the two most left characters.

The input in cell B2 is = LEFT (A2,2).

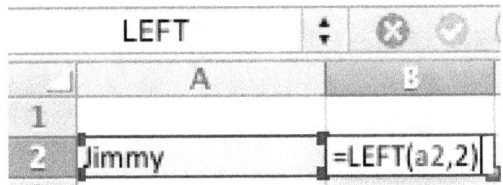

	A	B
1		
2	Jimmy	=LEFT(a2,2)

Now press enter. This action produces

Jimmy	Ji

Then we use the handle to produce

	A	B
1		
2	Jimmy	Ji
3	Jamie	Ja
4	Marsden	Ma
5	Oliver	Ol
6	Sean	Se
7	Jaxon	Ja
8	Bart	Ba
9	Daniel	Da
10	Dominic	Do
11	Alex	Al
12	Leo	Le
13	Vinu	Vi
14	Harry	Ha
15	Vincent	Vi
16	Oliver	Ol
17	Nathan	Na
18	Jean-Michael	Je
19	Scott	Sc
20	Matthew	Ma
21	Siddhartha	Si

The RIGHT function is used in an identical way.

The Upper Function is a simple function UPPER(string), which acts to convert all letters or characters in a string to upper case.

Here is an example of its use,
UPPER('cat') = CAT.
Here is the result of its use combined with the handle on a column of names.

B1		fx	=UPPER(A1)

	A	B	C	D
1	Pan	PAN		
2	Suen	SUEN		
3	Chandrakumar	CHANDRAKUMAR		
4	Lee	LEE		
5	Wu	WU		
6	Aitchison	AITCHISON		
7	Bell	BELL		
8	Hawthorne	HAWTHORNE		
9	Johnson	JOHNSON		
10	Phillips	PHILLIPS		
11	Vale	VALE		
12	Webber	WEBBER		
13	Brown-Bayliss	BROWN-BAYLISS		
14	Daffron	DAFFRON		
15	Beynon	BEYNON		

There are many more string functions in Excel but we will only look at one more, which is the MID function.

The structure or syntax of the function is MID(*string*, *number*1, *number*2).

What do these parameters mean?

The *string* is the string or word, which the function is acting on, *number*1 is the number of the letter in the string that you start with and *number*2 is the number of the letters in the string that you want, starting with the letter at *number*1.

Let's take the word, 'ELEPHANT' and calculate MID('ELEPHANT', 2, 5). The letter at position 2 is L and the fifth letter after starting with L is A, hence MID('ELEPHANT', 2, 5) = LEPHA.

Now let's use the MID function on a column of strings.

	B1						*fx*	=MID(A1, 1,2)

	A	B	C	D
1	Pan	Pa		
2	Suen	Su		
3	Chandrakumar	Ch		
4	Lee	Le		
5	Wu	Wu		
6	Aitchison	Ai		
7	Bell	Be		
8	Hawthorne	Ha		
9	Johnson	Jo		
10	Phillips	Ph		
11	Vale	Va		
12	Webber	We		
13	Brown-Bayliss	Br		
14	Daffron	Da		
15	Beynon	Be		

Chapter 10: OR

A widely used Excel function similar to the AND function is the OR function. It is either TRUE or FALSE depending on whether one or both of two conditions are true or not.

Here is an example, which illustrates this.

Example. Find the rows in which **either** the A and B cells are less than 50 in the table below.

	A	B
	E13	
1		
2	53	49
3	78	64
4	23	28
5	81	23
6	61	33
7	53	14
8	57	91
9	51	87
10	58	57
11	42	92

Into cell C2 type the function
declaration,' **=OR(A2<50,B2<50)**'
then

| 2 | 53 | 49 =OR(a2<50,b2<50) |

then press enter.

E14		↕ ⊗ ✓ ⌐ _fx_
A	B	C
53	49	TRUE
78	64	

Finally, complete the task by using the
handle.

	A	B	C
1			
2	53	49	TRUE
3	78	64	FALSE
4	23	28	TRUE
5	81	23	TRUE
6	61	33	TRUE
7	53	14	TRUE
8	57	91	FALSE
9	51	87	FALSE
10	58	57	FALSE
11	42	92	TRUE

In this case, we get TRUE for rows 2, 4, 5, 6, 7 and 11, as either one or both the A and B cells contain numbers, which are less than 50.

Structure or Syntax of OR

If we look at the structure or syntax of the OR function through a declaration, ' =OR(A2<50, B2<50)', we see the parameters of the OR function are OR(logical test, logical test). The parameters are two logical tests, the combination of whose results give different results.

Chapter 11: NOT

The NOT function is an interesting one. It only has a meaningful result if a cell has a data type called Booleans, which are either TRUE or FALSE. The number 1 is regarded as TRUE and 0 as FALSE. The function NOT gives a value of FALSE if the cell contains TRUE and TRUE if the cell contains FALSE.
Here is an example. In the last chapter, we got the table.

	A	B	C
1			
2	53	49	TRUE
3	78	64	FALSE
4	23	28	TRUE
5	81	23	TRUE
6	61	33	TRUE
7	53	14	TRUE
8	57	91	FALSE
9	51	87	FALSE
10	58	57	FALSE
11	42	92	TRUE

Use the NOT function on column C.

Into cell D2 type the function

declaration =**NOT(C2)**

| 2 | 53 | 49 | TRUE | =NOT(c2) |

then press enter.

A	B	C	D
53	49	TRUE	FALSE
78	64	FALSE	
23	28	TRUE	
81	23	TRUE	

Now finish the task by using the handle.

	A	B	C	D
1				
2	53	49	TRUE	FALSE
3	78	64	FALSE	TRUE
4	23	28	TRUE	FALSE
5	81	23	TRUE	FALSE
6	61	33	TRUE	FALSE
7	53	14	TRUE	FALSE
8	57	91	FALSE	TRUE
9	51	87	FALSE	TRUE
10	58	57	FALSE	TRUE
11	42	92	TRUE	FALSE

You may well ask what effect the NOT function has on cells, which contain data apart from Boolean. The table below gives the results on some such cells.

| | B9 | $\updownarrow$ | ⊗ | ⊘ | ⊙ | fx | =NOT(A9) |

	A	B	C	D
1				
2	1	FALSE		
3	0	TRUE		
4	3	FALSE		
5	cat	#VALUE!		
6	FALSE	TRUE		
7	45	FALSE		
8	23	FALSE		
9	117	FALSE		

It would seem any number except 0 is considered TRUE. Text like 'cat' is not recognized.

Only use the NOT function on Boolean data!

Chapter 12: XOR

A widely used Excel function similar to the OR function is the XOR function. Like the OR function, it is either TRUE or FALSE depending on whether one or both of two conditions are true or not. It is fussier than OR, as it is TRUE only if one of the conditions is TRUE. If both conditions are TRUE then the XOR function gives False.

Here is an Example. We want TRUE if **only one** of Family or First name has a length less than 7.

H	I
Family	**First**
Benton	Jimmy
Brooke	James
Cheong	Marshan
Griffith	Sam
Hicke	Sean
Howan	June
Jorgensen	Bob
McAndrew	Diana
Mildeng	Dominic
Pang	Alex

Into cell J2 type the function declaration,' =**XOR(LEN(H2)<7, LEN(I2)<7)** ' press enter then use the handle to get

Family	First	
Benton	Jimmy	FALSE
Brooke	James	FALSE
Cheong	Marshan	TRUE
Griffith	Sam	TRUE
Hicke	Sean	FALSE
Howan	June	FALSE
Jorgensen	Bob	TRUE
McAndrew	Diana	TRUE
Mildeng	Dominic	FALSE
Pang	Alex	FALSE

The reason that 'Benton Jimmy' gives FALSE is that the lengths of both 'Benton' and 'Jimmy' are less than 7 and with XOR this will give FALSE, whereas if we had used OR we would have got TRUE.

Structure or Syntax of XOR

We could look at the structure or syntax of the XOR function through a declaration, ' =**XOR(A2<50, B2<50)**' in the same way that we did for the AND and OR functions. If we do so, we see the parameters of the XOR function are XOR(logical test, logical test). The parameters are two logical tests; the combination of whose results give different results.

NB

The XOR function is a new one that was not in versions of Excel prior to 2013.

Chapter 13: SUMIF and SUMIFs

These functions are very similar to the SUM function except that they include the ability to put criteria in the function declaration. The best way to show this is with some examples.

Here is the first example. In the table of data below, we want to add all grades less than 50, using the functions of Excel.

H3		fx	
A	B	C	D
Name	First	Gender	Grade
Benson	Jane	F	66
Brook	James	M	64
Cheong	Bob	M	28
Griffin	Otis	M	23
Hicks	Sally	F	33
Howan	Jill	F	14
Jorgensen	Bob	M	91
McAndrew	Diana	F	87
Mildon	David	M	57
Pang	Alex	M	92
Suen	Luke	M	81
Chandrakum	Sakib	M	64
Lee	Helen	F	56
Wu	Valeri	F	78

Quite clearly the sum required is 28 + 23 + 33 + 14 = 98. Let's get this using Excel.

Into any cell, except those with the data, type the function declaration, ' =SUMIF(D2: D15,"<50", D2: D15) ' then press enter and you get 98.

Interestingly, as the range with the criterion < 50 is the same as the range we're summing over, we don't need the second D2: D13.

We would get exactly the same result if the function declaration was, ' =SUMIF(D2:D15,"<50") '. Try it and see for yourself.

If the range with the criterion is different from the range we're summing over this is not true.

Example. Find the sum of all grades of Female students.

A	B	C	D
Name	First	Gender	Grade
Benson	Jane	F	66
Brook	James	M	64
Cheong	Bob	M	28
Griffin	Otis	M	23
Hicks	Sally	F	33
Howan	Jill	F	14
Jorgensen	Bob	M	91
McAndrew	Diana	F	87
Mildon	David	M	57
Pang	Alex	M	92
Suen	Luke	M	81
Chandrakum	Sakib	M	64
Lee	Helen	F	56
Wu	Valeri	F	78

Into any cell, except those with the data, type the function declaration, ' =SUMIF(C2:C14,"F",D2:D15) ' then press enter. This time the result is 334.

The only trouble with SUMIF is that it only allows one criterion. If you want multiple criteria [plural of criterion] then you have to use SUMIFS. Once again an example will show what this entails.

Example. Find the sum of the grades of males whose school is 'A'.

	A	B	C	D	E
	Name	First	Gender	School	Grade
1	Benson	Jane	F	A	66
2	Brook	James	M	A	64
3	Cheong	Bob	M	B	28
4	Griffin	Otis	M	A	23
5	Hicks	Sally	F	B	33
6	Howan	Jill	F	A	14
7	Jorgensen	Bob	M	A	91
8	McAndrew	Diana	F	B	87
9	Mildon	David	M	B	57
10	Pang	Alex	M	B	92
11	Suen	Luke	M	A	81
12	Chandrakum	Sakib	M	B	64
13	Lee	Helen	F	B	56
14	Wu	Valeri	F	B	78

Into any cell, except those with the data, type the function declaration, ' =SUMIFS(E2:E15,C2:C15,"M", D2:D15,"A") ' then press enter. This time the result is 259.

The declaration is,' =SUMIFS(sum range, criterion 1, criterion 2, criterion 3,....)'. Sum range is always like E2: E15 and is the set you want to add over, criteria are always like D2: D15, "A", where you have a range where the criterion applies then a comma then the actual criterion.

These two functions have a structure or syntax so that the parameters are range and logical tests.

Chapter 14: COUNT and COUNTA

This chapter deals with two very simple functions: COUNT and COUNTA. All the COUNT function does is count the cells in a range, which contain numbers. The COUNTA function, however, counts all non-empty cells.

Example. Count all the cells with numbers only in the cells below.

M	N
Benson	McAndrew
56	Mildon
Cheong	Pang
45	Suen
78	33
Howan	90
Jorgensen	Wu
	7

Into any cell, except those with the data, type the function declaration, ' =COUNT(M2: M9, N2: N9) ' then press enter. The result is 6, as there six cells, which contain numbers.

Now let's use COUNTA. Into any cell, except those with the data, type the function declaration, ' =COUNTA(M2: M9, N2: N9) ' then press enter. The result is 15, as there fifteen cells, which contain numbers. Note that the empty cell is not counted, either by COUNT or COUNTA.

These two functions have a very simple structure or syntax. Their only parameter is a range.

Chapter 15: AVERAGEIF and AVERAGEIFs

The AVERAGEIF and AVERAGEIFS functions are very similar to the SUMIF and SUMIFS functions.

Example. Suppose you wanted to find the average grade of the female students in the table below.

	A	B	C	D	E
	K31			fx	
1	Name	First	Gender	School	Grade
2	Benson	Jane	F	A	66
3	Brook	James	M	A	64
4	Cheong	Bob	M	B	28
5	Griffin	Otis	M	A	23
6	Hicks	Sally	F	B	33
7	Howan	Jill	F	A	14
8	Jorgensen	Bob	M	A	91
9	McAndrew	Diana	F	B	87
10	Mildon	David	M	B	57
11	Pang	Alex	M	B	92
12	Suen	Luke	M	A	81
13	Chandrakum	Sakib	M	B	64
14	Lee	Helen	F	B	56
15	Wu	Valeri	F	B	78

Into any cell, except those with the data, type the function declaration, '
=AVERAGEIF(C2: C15, "F", E2: E15) '

then press enter. This time the result is
55.666....

Generally, the declaration is, '
=AVERAGEIF(criterion range, criterion,
average_range) '. The meaning of
average_range is the range you're going
to take the average of. AVERAGEIF is
for situations where there is only one
criterion.

If you have multiple criteria then you
must use AVERAGEIFS.

Example. Suppose you wanted to find
the average grade of the male students
who went to school B from the table
below.

	A	B	C	D	E
1	Name	First	Gender	School	Grade
2	Benson	Jane	F	A	66
3	Brook	James	M	A	64
4	Cheong	Bob	M	B	28
5	Griffin	Otis	M	A	23
6	Hicks	Sally	F	B	33
7	Howan	Jill	F	A	14
8	Jorgensen	Bob	M	A	91
9	McAndrew	Diana	F	B	87
10	Mildon	David	M	B	57
11	Pang	Alex	M	B	92
12	Suen	Luke	M	A	81
13	Chandrakum	Sakib	M	B	64
14	Lee	Helen	F	B	56
15	Wu	Valeri	F	B	78

Into any cell, except those with the data, type the function declaration, ' =AVERAGEIFS(E2:E15,C2:C15,"M", D2:D15,"B") ' then press enter. This time the result is 60.25

Generally, the declaration is, ' =AVERAGEIFs(average_range, criterion1 range, criterion1, criterion2 range, criterion2,) '. The meaning of average_range is the range you're going to take the average of. AVERAGEIFS is for situations where there is more than one criterion.

Chapter 16: LARGE and SMALL

The LARGE and SMALL functions are very easy to understand and use. The declaration ' = LARGE(F1: G100, 3)' finds the third largest number in the range F1: G100.

Example. Find the fifth largest number of the set in the cells below.

J	K
66	78
64	56
28	97
23	112
33	0.8
14	79
91	67
87	
57	
92	
81	
64	
56	
78	

Into any cell, except those with the data, type the function declaration, ' =LARGE(J2: K8, 5) ' then press enter. The result is 78.

Example. Find the second smallest
number of the set in the cells below.

J
66
64
28
23
33
14
91
87
57
92
81
64
56
78

Into any cell, except those with the data,
type the function declaration, '
=SMALL(J2: J15, 2) ' then press enter.
The result is 23.

Chapter 17: COUNTIF and COUNTIFS

The COUNTIF and COUNTIFS functions are very similar to the SUMIF and SUMIFS functions.

Example. Suppose you wanted to find female students were in the table below.

	K31	⬍	⊗ ⊘	fx	
	A	B	C	D	E
1	Name	First	Gender	School	Grade
2	Benson	Jane	F	A	66
3	Brook	James	M	A	64
4	Cheong	Bob	M	B	28
5	Griffin	Otis	M	A	23
6	Hicks	Sally	F	B	33
7	Howan	Jill	F	A	14
8	Jorgensen	Bob	M	A	91
9	McAndrew	Diana	F	B	87
10	Mildon	David	M	B	57
11	Pang	Alex	M	B	92
12	Suen	Luke	M	A	81
13	Chandrakum	Sakib	M	B	64
14	Lee	Helen	F	B	56
15	Wu	Valeri	F	B	78

Into any cell, except those with the data, type the function declaration, '

=COUNTIF(C2: C15, "F") ' then press enter. The result is 6

Generally, the declaration is, ' =COUNTIF(criterion range, criterion) '. COUNTIF is for situations where there is only one criterion.

If you have multiple criteria then you must use COUNTIFS.

Example. Suppose you wanted to find the number of male students who went to school B from the table below.

	A	B	C	D	E
	Name	First	Gender	School	Grade
1					
2	Benson	Jane	F	A	66
3	Brook	James	M	A	64
4	Cheong	Bob	M	B	28
5	Griffin	Otis	M	A	23
6	Hicks	Sally	F	B	33
7	Howan	Jill	F	A	14
8	Jorgensen	Bob	M	A	91
9	McAndrew	Diana	F	B	87
10	Mildon	David	M	B	57
11	Pang	Alex	M	B	92
12	Suen	Luke	M	A	81
13	Chandrakum	Sakib	M	B	64
14	Lee	Helen	F	B	56
15	Wu	Valeri	F	B	78
16					

Into any cell, except those with the data, type the function declaration, '
=COUNTIFS(C2:C15,"M", D2:D15,"B") '
then press enter. This time the result is
4.

Generally, the declaration is, '
=COUNTIFS(criterion1 range,
criterion1, criterion2 range,
criterion2,....) '. COUNTIFS is for
situations where there is more than one
criterion.

Chapter 18: VLOOKUP

The next two chapters deal with the VLOOKUP and HLOOKUP functions. These are very useful but are often misunderstood.

We will need a number of examples to try and make this clear. Here is the first example.

We have a table of grades from a school.

	A	B	C	D	E
	E36			fx	
	A	B	C	D	E
1	Family	First	Gender	English	Math
2	Argentin	Brady	M	56	71
3	Brewer	Shane	M	34	49
4	Jippu	Steven	M	80	88
5	Lattimore	Zelda	F	76	85
6	Liu	Phillippa	F	45	28
7	Martelil	Matthew	M	48	50
8	McTigure	Panama	F	65	57
9	Nichollson	Carla	F	95	88
10	Olsen	Brooklyn	F	48	53
11	Prapaithong	Jill	F	65	61
12	Ramez	Dillon	M	72	78
13	Sharo	William	M	35	41
14	Singh	Sakib	M	78	81
15	Stead	Helen	F	45	81
16	Stuart	Janine	F	56	61
17	Thompson	Mary	F	34	53
18	Wang	Maxine	F	80	57
19	Watson	Chloe	F	33	51
20	Wong	Vince	M	56	58
21	Zhang	Sue	F	67	42

and we want to put the math grade
beside each Family Name in the list
below.

H	I
Martelil	
Stuart	
Ramez	
Olsen	
McTigure	
Stead	
Singh	

This could be done by the extremely
tedious procedure of copying from the
table of grades to this table. If we did
this Martelil would have 50 beside him
and so on. Fortunately, the creators of
Excel foresaw this problem and devised
the LOOKUP functions.

Here is how this problem is solved using
the VLOOKUP function.
Into I2 type, ' =
VLOOKUP(H2,$A$2:$E$21, 5, FALSE) '
then press enter. The result is the value
50.

H	I
Martelil	50
Stuart	
Ramez	
Olsen	
McTigure	
Stead	
Singh	

We finish the procedure with the handles.

H	I
Martelil	50
Stuart	61
Ramez	78
Olsen	53
McTigure	57
Stead	81
Singh	81

Obviously, VLOOKUP works, but what does each part of VLOOKUP(H2,$A$2:$E$21, 5, FALSE) mean?

- H2 is the cell whose corresponding math grade is going to be looked up.

- $A$2:$E$21 is the range of values from which we will look up the math grade. Note the use of absolute references. This is usually **ESSENTIAL**!

- 5 refers to the column in the range where we get math grades. Column 1 is taken as the LOOKUP column. Its data must be in ASCENDING order.

- FALSE means we want an exact match between the names in our list and in the column Family. If we had used TRUE then the function would have been satisfied with an approximate match.

Finally, and these are extremely important:

1. **The column we are looking for matches in must be the first column in the range.**
2. **The first column must be put in ascending order.**

If you don't know how to order data then make sure to read the last chapter.

Now another example, which will hopefully help cement these ideas.

A company has 5 employees who are paid different hourly rates, as shown below in a Sheet1 of a spreadsheet using the range A1: B6.

	A	B
	A13	
1	Name	Hourly Rate
2	Alan	$40
3	Bob	$45
4	Cathy	$50
5	Diane	$35
6	Ed	$35

In order to pay these people, it is necessary to fill in the Hourly Rate in the table below in Sheet2 of the same spreadsheet.

	C15			⊗	⊘		*fx*

	A	B	C
1	Name	Hours	Hourly Rate
2	Diane	42	
3	Ed	36	
4	Alan	43	
5	Bob	29	
6	Cathy	56	
7			

We will use the VLOOKUP function to do this.

Into C2 type, ' = VLOOKUP(A2,Sheet1!$A$2:$B$6, 2, FALSE) ' then press enter. Sheet1! tells VLOOKUP to go to Sheet1 to find the range $A$2:$B$6. The result is the value $35 then use the handle to complete the table.

	G37			⊗	⊘		*fx*

	A	B	C
1	Name	Hours	Hourly Rate
2	Diane	42	$35
3	Ed	36	$35
4	Alan	43	$40
5	Bob	29	$45
6	Cathy	56	$50

Before we began it was necessary to make sure that a check was made that the data in the lookup column was in ascending order. As it was, there was no problem.

We can easily calculate the week's pay of the employees by typing, '= B2*C2', into D2 then pressing enter followed by the handle.

| G8 | | | | | fx | |

	A	B	C	D
1	Name	Hours	Hourly Rate	Week's Pay
2	Diane	42	$35	$1,470
3	Ed	36	$35	$1,260
4	Alan	43	$40	$1,720
5	Bob	29	$45	$1,305
6	Cathy	56	$50	$2,800

Before finishing this function, let us examine the last parameter, which takes the value FALSE for an exact match and TRUE for an approximate match.
Leave the function as it is but change Diane to Diana. The table below shows what happens.

	C9			⊗	✓		*fx*	

	A	B	C	D
1	**Name**	**Hours**	**Hourly Rate**	**Week's Pay**
2	Diana	42	#N/A	#N/A
3	Ed	36	$35	$1,260
4	Alan	43	$40	$1,720
5	Bob	29	$45	$1,305
6	Cathy	56	$50	$2,800

However, now change FALSE to TRUE in the function declaration, ' = VLOOKUP(A2, Sheet1!$A$2:$B$6, 2, TRUE) ' and then press enter followed by the handle.

	C12			⊗	✓		*fx*	

	A	B	C	D
1	**Name**	**Hours**	**Hourly Rate**	**Week's Pay**
2	Diana	42	$50	$2,100
3	Ed	36	$35	$1,260
4	Alan	43	$40	$1,720
5	Bob	29	$45	$1,305
6	Cathy	56	$50	$2,800

VLOOKUP has assigned Cathy's hourly rate, probably because the names Cathy and Diana have the same number of characters or letters. Cathy and Diana are regarded as approximately equal.

Finally, we have a look at the syntax of the VLOOKUP function. As usual, a declaration is very useful: ' = VLOOKUP(H2,$A$2:$E$21, 5, FALSE) '.

The parameters in order are, a cell [where the value to be looked up is], the range where the function looks, the number of the column where the value is found and a Boolean value of TRUE or FALSE.

Chapter 19: HLOOKUP

VLOOKUP used data arranged
vertically; HLOOKUP uses data
arranged horizontally.

How can data be arranged horizontally?
Here is an example:

	A	B	C	D	E
A8				fx	
1	Make	Ford	Toyota	Honda	Chrysler
2	In Stock	20	23	11	5
3	Last Sale	May 3,2019	May 2, 2019	May 3, 2019	May 1, 2019

Here is how you could access this simple
table using HLOOKUP. The problem is
to find the number of Honda cars in
stock, using the HLOOKUP function,
and write it in the cell B6 to the right of
the word Honda in cell A6.

6	Honda	

In cell B6, type the function declaration,'
=HLOOKUP("Honda",A1:E3, 2, FALSE)'
then press enter.

6	Honda	11

We get the same result if we have '
=HLOOKUP(A6, A1: E3, 2, FALSE)'.
The reason is that the word 'Honda' is
contained in A6.

However, see what happens if have '
=HLOOKUP(A2, A1: E3, 2, FALSE)'.

6	Honda	#N/A

The reason is that the word 'Honda' is
not contained in A2.

Now, see what happens if have '
=HLOOKUP("FORD", A1: E3, 2,
FALSE)'.

6	Honda	20

The reason is that the word 'Ford' is
contained in the top row of the table
with 20 below it in row 2.

As with VLOOKUP, we need to examine
the function declaration and explain
what is happening. Here is the original
function declaration, '

=HLOOKUP("Honda", A1:E3, 2, FALSE)'.

HLOOKUP searches for the word Honda in the **top row** of the range A1: E3. Once this is found, the 2 tells HLOOKUP to take the contents of row 2, which is beneath Honda and put it in B6. The FALSE tells HLOOKUP that there must be an exact match. An approximate match, such as HINDA or HONDO would not be acceptable, whereas it might for TRUE.

Finally, you may recall that VLOOKUP required the left column to be in ascending order. For HLOOKUP, this is not necessary for a last parameter value of FALSE. However, it is necessary that the top row of the range be in ascending order if the last parameter is TRUE.

We have often finished chapters on functions with a paragraph about structure or syntax. We will not dwell on structure or syntax. The syntax is very similar to that for VLOOKUP. If you are interested then see what was written in that chapter about this topic.

Chapter 20: A few notes about Pasting, Ordering and Filtering

Usually, basic information such as this is covered early. However, the notes about ordering are only vital when you use LOOKUP functions.

Pasting

Here is the result of multiplying two columns of numbers using the declaration, **'= A2*B2'** then the handle.

C2				fx	=A2*B2
	A	B	C	D	
1					
2	3	4	12		
3	7	9	63		
4	5	6	30		
5	4	5	20		
6	6	9	54		

Often having computed a column like this it might be useful to eliminate one or both of the source columns but if you delete them this is what happens.

C2					fx	=#REF!*B2
	A	B		C		D
1						
2			4	#REF!		
3			9	#REF!		
4			6	#REF!		
5			5	#REF!		
6			9	#REF!		

To avoid this use Paste Values

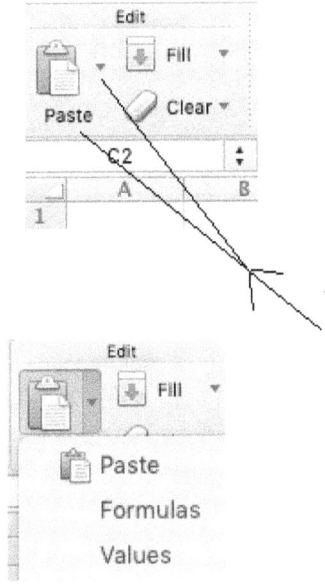

after copying the original.

C2				fx	12
	A	B	C		D
1					
2			4	12	
3			9	63	
4			6	30	
5			5	20	
6			9	54	

Now after deleting one the table of products is unaffected, as can be seen by looking at the diagram above.

Ordering

As mentioned and emphasized in the chapter on VLOOKUP, it is absolutely important that the first column is in ascending order. This is all well and good but how is this done?

Suppose the range we are going to look up from is the table below.

H21				fx
	A	B	C	D	E
	ID	Family	First	Gender	Wage per hour
1	ID	Family	First	Gender	Wage per hour
2	B45	Benton	Jimmy	M	$25
3	G06	Brooke	Helen	F	$25
4	A56	Cheong	Greta	F	$23
5	H67	Griffith	Murgatroyd	M	$30
6	A01	Hicke	Sue	F	$25
7	A12	Howan	June	F	$67
8	H02	Jorgensen	Fred	M	$32
9	H13	McAndrew	Diana	F	$45
10	G22	Mildeng	Zane	M	$56
11	F11	Pang	Peter	M	$25
12					

and suppose the lookup is based on ID. You can see that the table has been ordered by Family.

To order this range by ID is very easy.

Here are the steps, after you have selected the range A2: E11.

| A2 | | | | fx | B45 |

	A	B	C	D	E
	ID	Family	First	Gender	Wage per hour
1	ID	Family	First	Gender	Wage per hour
2	B45	Benton	Jimmy	M	$25
3	G06	Brooke	Helen	F	$25
4	A56	Cheong	Greta	F	$23
5	H67	Griffith	Murgatroyd	M	$30
6	A01	Hicke	Sue	F	$25
7	A12	Howan	June	F	$67
8	H02	Jorgensen	Fred	M	$32
9	H13	McAndrew	Diana	F	$45
10	G22	Mildeng	Zane	M	$56
11	F11	Pang	Peter	M	$25

(1) Click on the DATA tab on your Excel Menu.

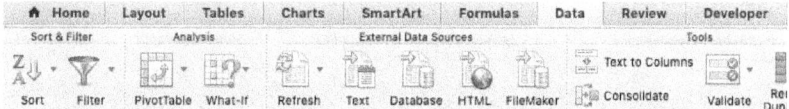

(2) Open the Sort & Filter window.

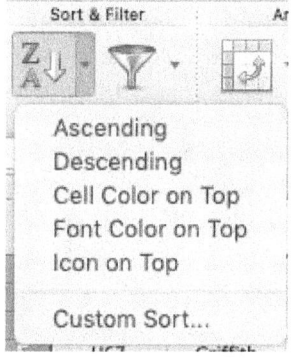

89

(3) Pick Custom Sort [I always use this except for trivial sorts]

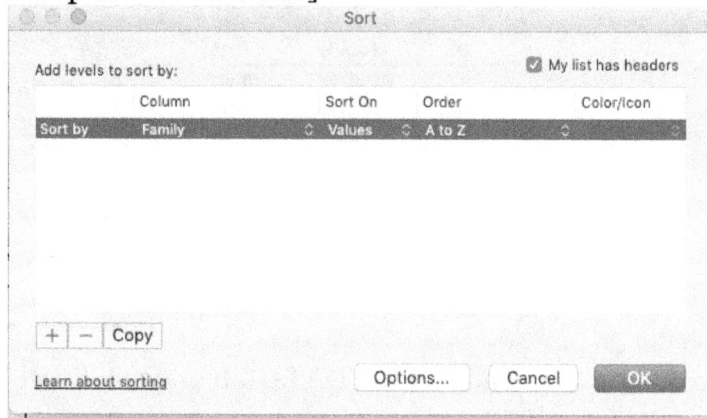

(4) Click on Family

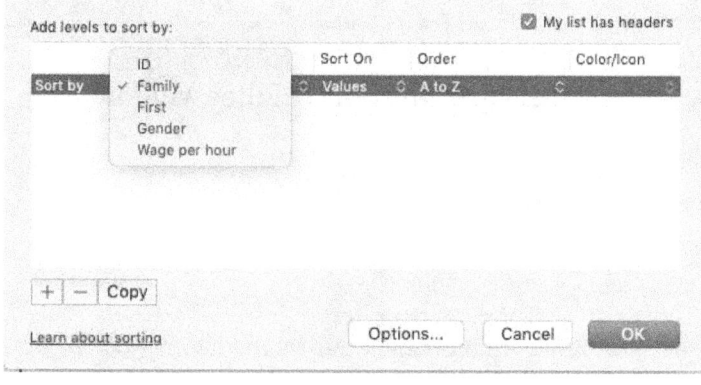

(5) Select ID and press OK.

	A	B	C	D	E
	A2			fx	A01
1	ID	Family	First	Gender	Wage per hour
2	A01	Hicke	Sue	F	$25
3	A12	Howan	June	F	$67
4	A56	Cheong	Greta	F	$23
5	B45	Benton	Jimmy	M	$25
6	F11	Pang	Peter	M	$25
7	G06	Brooke	Helen	F	$25
8	G22	Mildeng	Zane	M	$56
9	H02	Jorgensen	Fred	M	$32
10	H13	McAndrew	Diana	F	$45
11	H67	Griffith	Murgatroyd	M	$30

The VLOOKUP function can now be used.

Further Ordering

Consider the following data set of student results from a high school.

	A	B	C	D	E	F	G
1	ID	Family	First	Gender	English	Math	Science
2	A67	Argentine	Brad	M	47	64	88
3	M78	Chadhaman	Dipsya	F	61	42	57
4	D57	Davison	Aimee	F	61	57	50
5	A98	Gleeson	Wallace	M	50	28	42
6	F56	Greenwell	Fiona	F	53	92	57
7	B01	Hrstich	Jack	M	57	23	57
8	Y67	Huang	Stephani	F	28	57	51
9	A21	Jitney	Steve	M	59	66	46
10	C34	Kennedy	Mary	F	81	87	47
11	F82	Leotard	Tereise	F	57	81	53
12	Y76	Li	Jill	F	88	77	58
13	B98	Motlagh	Sam	M	78	14	71
14	F83	Pecovera	Ellen	F	51	64	78
15	H25	Randal	Sophie	F	49	46	61
16	H03	Raufger	Shazinat	F	58	56	81
17	A09	Shao	Bill	M	71	45	78
18	Z07	Sidwell	Amelia	F	45	71	42
19	B67	Smith	Jos	M	53	33	77
20	H46	Sproully	Larissa	F	85	88	53
21	H23	Sung	Kate	F	42	78	81
22	C02	Zumpar	Don	M	31	21	57

We want the results of the female students only with a total over the three subjects and the results put in descending order, with the highest total at the top, down to the lowest.

First, let's get the totals.
Into H2, type the formula declaration, '**= E2 + F2 + G2**' then press equal. Once this is done use the handle to get the data set below.

	A	B	C	D	E	F	G	H
1	ID	Family	First	Gender	English	Math	Science	Total
2	A67	Argentine	Brad	M	47	64	88	199
3	M78	Chadhaman	Dipsya	F	61	42	57	160
4	D57	Davison	Aimee	F	61	57	50	168
5	A98	Gleeson	Wallace	M	50	28	42	120
6	F56	Greenwell	Fiona	F	53	92	57	202
7	B01	Hrstich	Jack	M	57	23	57	137
8	Y67	Huang	Stephani	F	28	57	51	136
9	A21	Jitney	Steve	M	59	66	46	171
10	C34	Kennedy	Mary	F	81	87	47	215
11	F82	Leotard	Tereise	F	57	81	53	191
12	Y76	Li	Jill	F	88	77	58	223
13	B98	Motlagh	Sam	M	78	14	71	163
14	F83	Pecovera	Ellen	F	51	64	78	193
15	H25	Randal	Sophie	F	49	46	61	156
16	H03	Raufger	Shazinat	F	58	56	81	195
17	A09	Shao	Bill	M	71	45	78	194
18	Z07	Sidwell	Amelia	F	45	71	42	158
19	B67	Smith	Jos	M	53	33	77	163
20	H46	Sproully	Larissa	F	85	88	53	226
21	H23	Sung	Kate	F	42	78	81	201
22	C02	Zumpar	Don	M	31	21	57	109

Now select A1: H22 and go to Data.

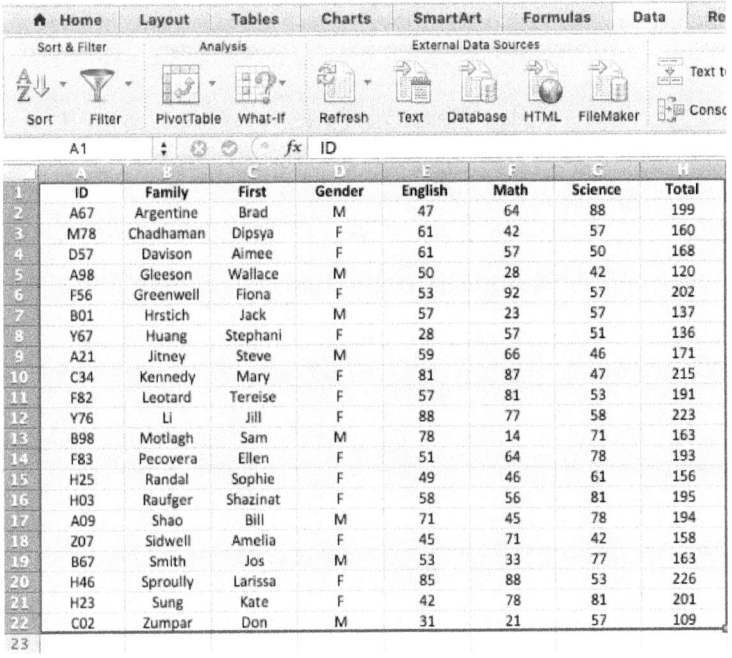

	A	B	C	D	E	F	G	H
1	ID	Family	First	Gender	English	Math	Science	Total
2	A67	Argentine	Brad	M	47	64	88	199
3	M78	Chadhaman	Dipsya	F	61	42	57	160
4	D57	Davison	Aimee	F	61	57	50	168
5	A98	Gleeson	Wallace	M	50	28	42	120
6	F56	Greenwell	Fiona	F	53	92	57	202
7	B01	Hrstich	Jack	M	57	23	57	137
8	Y67	Huang	Stephani	F	28	57	51	136
9	A21	Jitney	Steve	M	59	66	46	171
10	C34	Kennedy	Mary	F	81	87	47	215
11	F82	Leotard	Tereise	F	57	81	53	191
12	Y76	Li	Jill	F	88	77	58	223
13	B98	Motlagh	Sam	M	78	14	71	163
14	F83	Pecovera	Ellen	F	51	64	78	193
15	H25	Randal	Sophie	F	49	46	61	156
16	H03	Raufger	Shazinat	F	58	56	81	195
17	A09	Shao	Bill	M	71	45	78	194
18	Z07	Sidwell	Amelia	F	45	71	42	158
19	B67	Smith	Jos	M	53	33	77	163
20	H46	Sproully	Larissa	F	85	88	53	226
21	H23	Sung	Kate	F	42	78	81	201
22	C02	Zumpar	Don	M	31	21	57	109
23								

Go to Sort & Filter and do a custom sort based on Gender. Here is what you should have.

	A	B	C	D	E	F	G	H
	ID	Family	First	Gender	English	Math	Science	Total
1	ID	Family	First	Gender	English	Math	Science	Total
2	M78	Chadhaman	Dipsya	F	61	42	57	160
3	D57	Davison	Aimee	F	61	57	50	168
4	F56	Greenwell	Fiona	F	53	92	57	202
5	Y67	Huang	Stephani	F	28	57	51	136
6	C34	Kennedy	Mary	F	81	87	47	215
7	F82	Leotard	Tereise	F	57	81	53	191
8	Y76	Li	Jill	F	88	77	58	223
9	F83	Pecovera	Ellen	F	51	64	78	193
10	H25	Randal	Sophie	F	49	46	61	156
11	H03	Raufger	Shazinat	F	58	56	81	195
12	Z07	Sidwell	Amelia	F	45	71	42	158
13	H46	Sproully	Larissa	F	85	88	53	226
14	H23	Sung	Kate	F	42	78	81	201
15	A67	Argentine	Brad	M	47	64	88	199
16	A98	Gleeson	Wallace	M	50	28	42	120
17	B01	Hrstich	Jack	M	57	23	57	137
18	A21	Jitney	Steve	M	59	66	46	171
19	B98	Motlagh	Sam	M	78	14	71	163
20	A09	Shao	Bill	M	71	45	78	194
21	B67	Smith	Jos	M	53	33	77	163
22	C02	Zumpar	Don	M	31	21	57	109

Now select the range A1: H14, which is the range with females and copy this range [ctrl-c in a PC or command-c in a Mac]. Paste in some place, using a special paste with Values. The result is shown below.

ID	Family	First	Gender	English	Math	Science	Total
M78	Chadhaman	Dipsya	F	61	42	57	160
D57	Davison	Aimee	F	61	57	50	168
F56	Greenwell	Fiona	F	53	92	57	202
Y67	Huang	Stephani	F	28	57	51	136
C34	Kennedy	Mary	F	81	87	47	215
F82	Leotard	Tereise	F	57	81	53	191
Y76	Li	Jill	F	88	77	58	223
F83	Pecovera	Ellen	F	51	64	78	193
H25	Randal	Sophie	F	49	46	61	156
H03	Raufger	Shazinat	F	58	56	81	195
Z07	Sidwell	Amelia	F	45	71	42	158
H46	Sproully	Larissa	F	85	88	53	226
H23	Sung	Kate	F	42	78	81	201

Now we need to sort them via total in descending order. Once again select the range and go to Sort and Filter. Select Custom Sort and make sure you pick Total and Largest to Smallest.

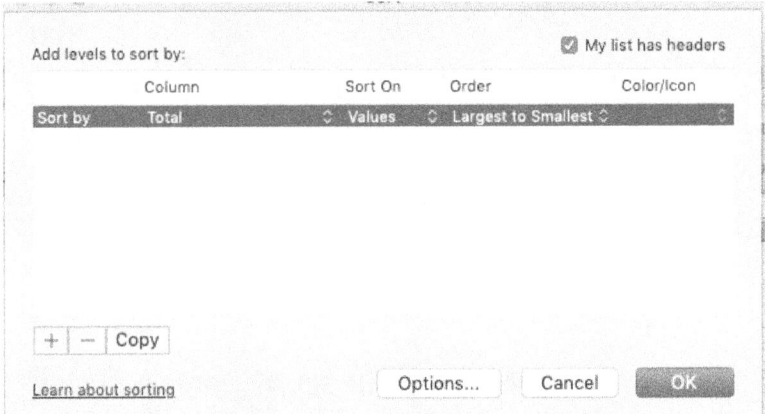

We now have the desired table.

ID	Family	First	Gender	English	Math	Science	Total
H46	Sproully	Larissa	F	85	88	53	226
Y76	Li	Jill	F	88	77	58	223
C34	Kennedy	Mary	F	81	87	47	215
F56	Greenwell	Fiona	F	53	92	57	202
H23	Sung	Kate	F	42	78	81	201
H03	Raufger	Shazinat	F	58	56	81	195
F83	Pecovera	Ellen	F	51	64	78	193
F82	Leotard	Tereise	F	57	81	53	191
D57	Davison	Aimee	F	61	57	50	168
M78	Chadhaman	Dipsya	F	61	42	57	160
Z07	Sidwell	Amelia	F	45	71	42	158
H25	Randal	Sophie	F	49	46	61	156
Y67	Huang	Stephani	F	28	57	51	136

Some of you may have wondered about
Filter. What is this? Read on to find out.

Filter

I will not spend too much time on this.
Basically, it is a way of examining a set
of data using constraints but leaving the
data set intact.

Here is an example. The data set below
is the same one we extracted the females
from before using order. This time we
are going to have a look at the male
results in ascending order of Total
[smallest to largest], save the filtered

data set elsewhere then restore the data set.

	A	B	C	D	E	F	G	H
M28			fx					
	A	B	C	D	E	F	G	H
1	ID	Family	First	Gender	English	Math	Science	Total
2	H46	Sproully	Larissa	F	85	88	53	226
3	Y76	Li	Jill	F	88	77	58	223
4	C34	Kennedy	Mary	F	81	87	47	215
5	F56	Greenwell	Fiona	F	53	92	57	202
6	H23	Sung	Kate	F	42	78	81	201
7	H03	Raufger	Shazinat	F	58	56	81	195
8	A09	Shao	Bill	M	71	45	78	194
9	F83	Pecovera	Ellen	F	51	64	78	193
10	F82	Leotard	Tereise	F	57	81	53	191
11	A21	Jitney	Steve	M	59	66	46	171
12	D57	Davison	Aimee	F	61	57	50	168
13	B98	Motlagh	Sam	M	78	14	71	163
14	B67	Smith	Jos	M	53	33	77	163
15	M78	Chadhaman	Dipsya	F	61	42	57	160
16	Z07	Sidwell	Amelia	F	45	71	42	158
17	A67	Argentine	Brad	M	47	64	88	199
18	H25	Randal	Sophie	F	49	46	61	156
19	B01	Hrstich	Jack	M	57	23	57	137
20	Y67	Huang	Stephani	F	28	57	51	136
21	A98	Gleeson	Wallace	M	50	28	42	120
22	C02	Zumpar	Don	M	31	21	57	109
23								
24								

Go to the Sort & Filter tab, as you did before, after clicking on the Data tab of the Excel menu.

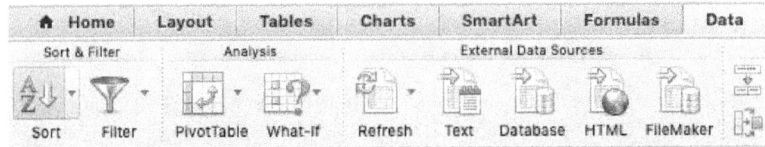

This time click Filter, but NOT the down arrow, after selecting the whole data set.

	A	B	C	D	E	F	G	H
1	ID ▼	Family ▼	First ▼	Gender ▼	English ▼	Math ▼	Science ▼	Total ▼
2	H46	Sproully	Larissa	F	85	88	53	226
3	Y76	Li	Jill	F	88	77	58	223
4	C34	Kennedy	Mary	F	81	87	47	215
5	F56	Greenwell	Fiona	F	53	92	57	202
6	H23	Sung	Kate	F	42	78	81	201
7	H03	Raufger	Shazinat	F	58	56	81	195
8	A09	Shao	Bill	M	71	45	78	194
9	F83	Pecovera	Ellen	F	51	64	78	193
10	F82	Leotard	Tereise	F	57	81	53	191
11	A21	Jitney	Steve	M	59	66	46	171
12	D57	Davison	Aimee	F	61	57	50	168
13	B98	Motlagh	Sam	M	78	14	71	163
14	B67	Smith	Jos	M	53	33	77	163
15	M78	Chadhaman	Dipsya	F	61	42	57	160
16	Z07	Sidwell	Amelia	F	45	71	42	158
17	A67	Argentine	Brad	M	47	64	88	199
18	H25	Randal	Sophie	F	49	46	61	156
19	B01	Hrstich	Jack	M	57	23	57	137
20	Y67	Huang	Stephani	F	28	57	51	136
21	A98	Gleeson	Wallace	M	50	28	42	120
22	C02	Zumpar	Don	M	31	21	57	109
23								
24								

You will notice little arrows. Click on the arrow in the Gender column. Immediately you do, the following window appears.

Fill it in as shown.

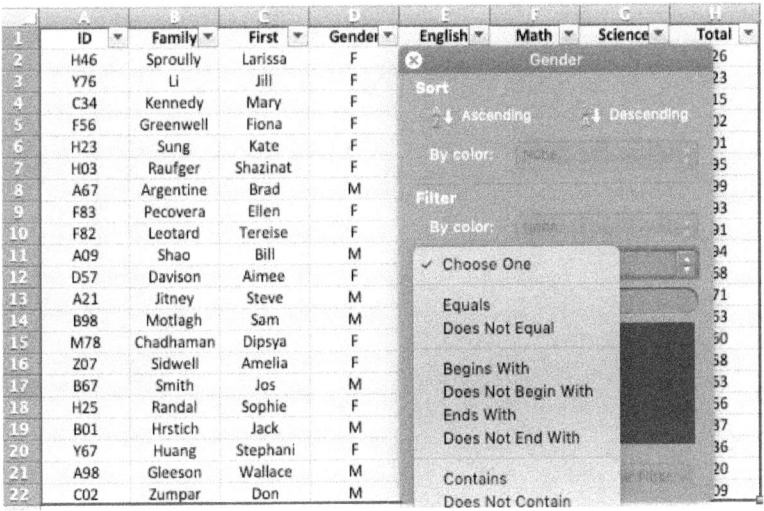

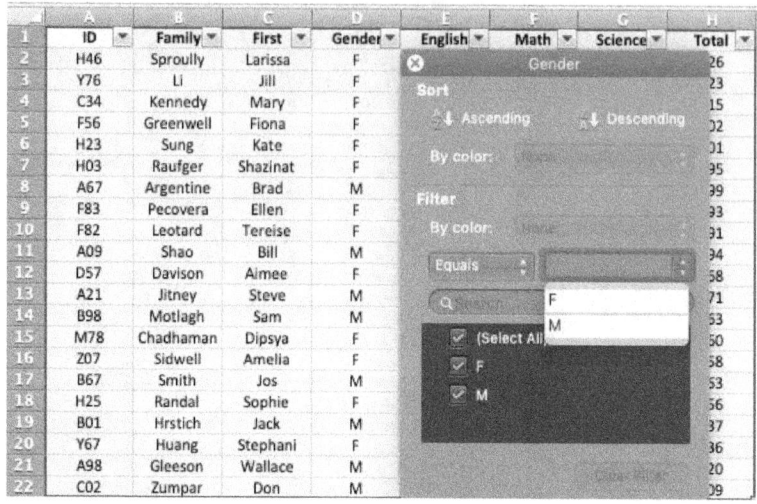

Leading to

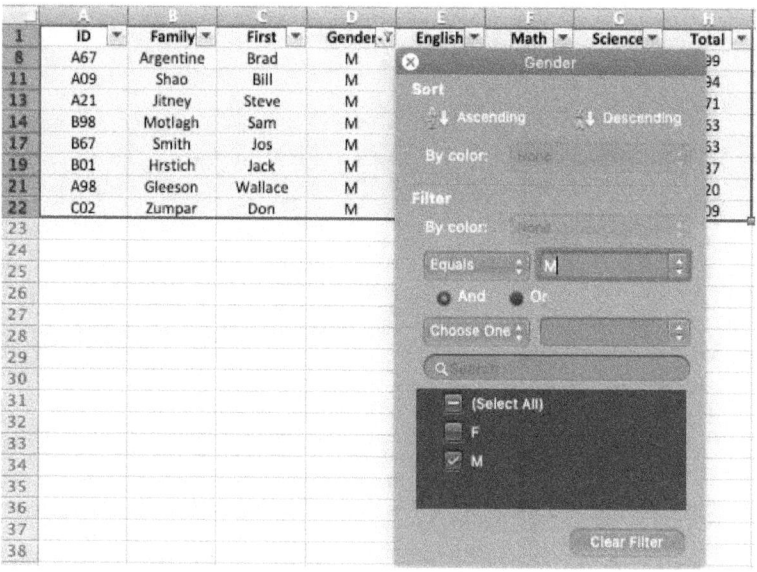

Finally, close the Gender window and you only have the Male results.

	A	B	C	D	E	F	G	H
1	ID ▼	Family ▼	First ▼	Gender ▼	English ▼	Math ▼	Science ▼	Total ▼
8	A67	Argentine	Brad	M	47	64	88	199
11	A09	Shao	Bill	M	71	45	78	194
13	A21	Jitney	Steve	M	59	66	46	171
14	B98	Motlagh	Sam	M	78	14	71	163
17	B67	Smith	Jos	M	53	33	77	163
19	B01	Hrstich	Jack	M	57	23	57	137
21	A98	Gleeson	Wallace	M	50	28	42	120
22	C02	Zumpar	Don	M	31	21	57	109

These results are in descending order.
However, as it is necessary to put them
in ascending order, you use the little
arrow beside Total. A little window
appears in which ascending is an option.

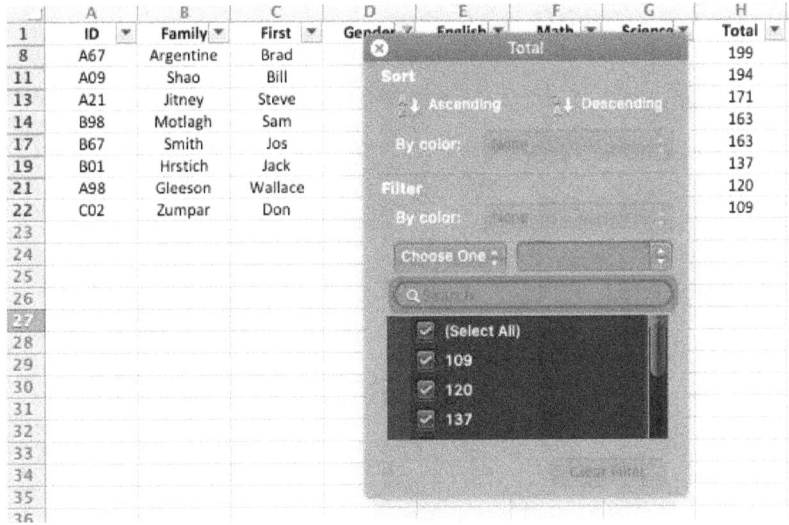

Clicking on this will put the results in
ascending order.

	A	B	C	D	E	F	G	H
	ID ▼	Family ▼	First ▼	Gender ▼	English ▼	Math ▼	Science ▼	Total ▼
1								
8	C02	Zumpar	Don	M	31	21	57	109
11	A98	Gleeson	Wallace	M	50	28	42	120
13	B01	Hrstich	Jack	M	57	23	57	137
14	B98	Motlagh	Sam	M	78	14	71	163
17	B67	Smith	Jos	M	53	33	77	163
19	A21	Jitney	Steve	M	59	66	46	171
21	A09	Shao	Bill	M	71	45	78	194
22	A67	Argentine	Brad	M	47	64	88	199

If you want to save this as a data set then just select the set and paste it somewhere. I have pasted it just below the filter. Notice that the data is copied as Values.

	A	B	C	D	E	F	G	H
	ID ▼	Family ▼	First ▼	Gender ▼	English ▼	Math ▼	Science ▼	Total ▼
1								
8	C02	Zumpar	Don	M	31	21	57	109
11	A98	Gleeson	Wallace	M	50	28	42	120
13	B01	Hrstich	Jack	M	57	23	57	137
14	B98	Motlagh	Sam	M	78	14	71	163
17	B67	Smith	Jos	M	53	33	77	163
19	A21	Jitney	Steve	M	59	66	46	171
21	A09	Shao	Bill	M	71	45	78	194
22	A67	Argentine	Brad	M	47	64	88	199
23								
24	ID	Family	First	Gender	English	Math	Science	Total
25	C02	Zumpar	Don	M	31	21	57	109
26	A98	Gleeson	Wallace	M	50	28	42	120
27	B01	Hrstich	Jack	M	57	23	57	137
28	B98	Motlagh	Sam	M	78	14	71	163
29	B67	Smith	Jos	M	53	33	77	163
30	A21	Jitney	Steve	M	59	66	46	171
31	A09	Shao	Bill	M	71	45	78	194
32	A67	Argentine	Brad	M	47	64	88	199

Finally, restoring the original set of data is easy. Just click on the Filter icon. When you do this the Filter arrows are removed and you have the original data set restored.

	A	B	C	D	E	F	G	H
1	ID	Family	First	Gender	English	Math	Science	Total
2	H46	Sproully	Larissa	F	85	88	53	226
3	Y76	Li	Jill	F	88	77	58	223
4	C34	Kennedy	Mary	F	81	87	47	215
5	F56	Greenwell	Fiona	F	53	92	57	202
6	H23	Sung	Kate	F	42	78	81	201
7	H03	Raufger	Shazinat	F	58	56	81	195
8	C02	Zumpar	Don	M	31	21	57	109
9	F83	Pecovera	Ellen	F	51	64	78	193
10	F82	Leotard	Tereise	F	57	81	53	191
11	A98	Gleeson	Wallace	M	50	28	42	120
12	D57	Davison	Aimee	F	61	57	50	168
13	B01	Hrstich	Jack	M	57	23	57	137
14	B98	Motlagh	Sam	M	78	14	71	163
15	M78	Chadhaman	Dipsya	F	61	42	57	160
16	Z07	Sidwell	Amelia	F	45	71	42	158
17	B67	Smith	Jos	M	53	33	77	163
18	H25	Randal	Sophie	F	49	46	61	156
19	A21	Jitney	Steve	M	59	66	46	171
20	Y67	Huang	Stephani	F	28	57	51	136
21	A09	Shao	Bill	M	71	45	78	194
22	A67	Argentine	Brad	M	47	64	88	199
23								

Filtering by Color

While exploring Filter, you may have noticed the option Filter by Cell Color.

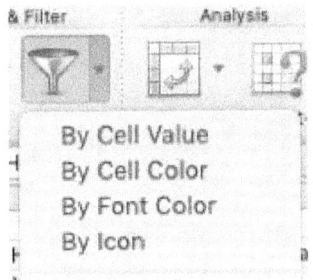

Hitherto, this has been irrelevant, as we have only used Black and White. However, Excel offers all sorts of color

schemes. Two simple ones for Fill Color
and Font Color are shown.

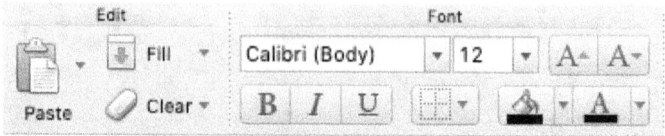

To show the use of Filter by Cell Color,
we have colored the data set we have
been using.

	A	B	C	D	E	F	G	H
1	ID	Family	First	Gender	English	Math	Science	Total
2	H46	Sproully	Larissa	F	85	88	53	226
3	Y76	Li	Jill	F	88	77	58	223
4	C34	Kennedy	Mary	F	81	87	47	215
5	F56	Greenwell	Fiona	F	53	92	57	202
6	H23	Sung	Kate	F	42	78	81	201
7	H03	Raufger	Shazinat	F	58	56	81	195
8	C02	Zumpar	Don	M	31	21	57	109
9	F83	Pecovera	Ellen	F	51	64	78	193
10	F82	Leotard	Tereise	F	57	81	53	191
11	A98	Gleeson	Wallace	M	50	28	42	120
12	D57	Davison	Aimee	F	61	57	50	168
13	B01	Hrstich	Jack	M	57	23	57	137
14	B98	Motlagh	Sam	M	78	14	71	163
15	M78	Chadhaman	Dipsya	F	61	42	57	160
16	Z07	Sidwell	Amelia	F	45	71	42	158
17	B67	Smith	Jos	M	53	33	77	163
18	H25	Randal	Sophie	F	49	46	61	156
19	A21	Jitney	Steve	M	59	66	46	171
20	Y67	Huang	Stephani	F	28	57	51	136
21	A09	Shao	Bill	M	71	45	78	194
22	A67	Argentine	Brad	M	47	64	88	199

Now click on Filter after selecting the
data set. This results in the following.

	A	B	C	D	E	F	G	H
	ID	Family	First	Gender	English	Math	Science	Total
1	ID	Family	First	Gender	English	Math	Science	Total
2	H46	Sproully	Larissa	F	85	88	53	226
3	Y76	Li	Jill	F	88	77	58	223
4	C34	Kennedy	Mary	F	81	87	47	215
5	F56	Greenwell	Fiona	F	53	92	57	202
6	H23	Sung	Kate	F	42	78	81	201
7	H03	Raufger	Shazinat	F	58	56	81	195
8	C02	Zumpar	Don	M	31	21	57	109
9	F83	Pecovera	Ellen	F	51	64	78	193
10	F82	Leotard	Tereise	F	57	81	53	191
11	A98	Gleeson	Wallace	M	50	28	42	120
12	D57	Davison	Aimee	F	61	57	50	168
13	B01	Hrstich	Jack	M	57	23	57	137
14	B98	Motlagh	Sam	M	78	14	71	163
15	M78	Chadhaman	Dipsya	F	61	42	57	160
16	Z07	Sidwell	Amelia	F	45	71	42	158
17	B67	Smith	Jos	M	53	33	77	163
18	H25	Randal	Sophie	F	49	46	61	156
19	A21	Jitney	Steve	M	59	66	46	171
20	Y67	Huang	Stephani	F	28	57	51	136
21	A09	Shao	Bill	M	71	45	78	194
22	A67	Argentine	Brad	M	47	64	88	199
23								

Now use any little arrow, choose color

and finally yellow or red.

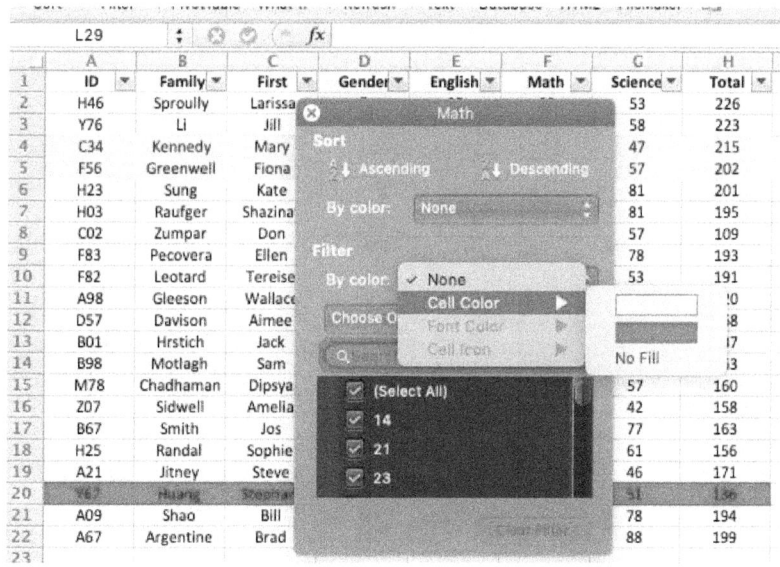

If we pick red then we get the filter below.

	A	B	C	D	E	F	G	H
	ID ▼	Family ▼	First ▼	Gender ▼	English ▼	Math ⋅▼	Science ▼	Total ▼
20	Y67	Huang	Stephani	F	28	57	51	136

We restore the original by clicking on the Filter icon.

	A	B	C	D	E	F	G	H
1	ID	Family	First	Gender	English	Math	Science	Total
2	H46	Sproully	Larissa	F	85	88	53	226
3	Y76	Li	Jill	F	88	77	58	223
4	C34	Kennedy	Mary	F	81	87	47	215
5	F56	Greenwell	Fiona	F	53	92	57	202
6	H23	Sung	Kate	F	42	78	81	201
7	H03	Raufger	Shazinat	F	58	56	81	195
8	C02	Zumpar	Don	M	31	21	57	109
9	F83	Pecovera	Ellen	F	51	64	78	193
10	F82	Leotard	Tereise	F	57	81	53	191
11	A98	Gleeson	Wallace	M	50	28	42	120
12	D57	Davison	Aimee	F	61	57	50	168
13	B01	Hrstich	Jack	M	57	23	57	137
14	B98	Motlagh	Sam	M	78	14	71	163
15	M78	Chadhaman	Dipsya	F	61	42	57	160
16	Z07	Sidwell	Amelia	F	45	71	42	158
17	B67	Smith	Jos	M	53	33	77	163
18	H25	Randal	Sophie	F	49	46	61	156
19	A21	Jitney	Steve	M	59	66	46	171
20	Y67	Huang	Stephani	F	28	57	51	136
21	A09	Shao	Bill	M	71	45	78	194
22	A67	Argentine	Brad	M	47	64	88	199

Conclusion

Excel has a computer language called VBA [Visual Basic for Applications] associated with it. Really good exponents of Excel use VBA all the time

If you wish to master it you must know the basics of formulas and functions inside and out.

This book has thoroughly covered many of those basics.

Before you proceed further with formulas and functions you must master the material in the book to the extent that you can do the problems in every chapter.

Once you have mastered the material in this book you are ready for all the other things that can be done with EXCEL and the VBA programming language!

Good Luck.

Welcome to the last page reader, I'm
happy to see you here I hope you had a
great time reading my book and if you
want to support my work, leaving an
honest review will be highly appreciated.
Thank you so much!

Respectfully,
William B. Skates